GAO
Accountability * Integrity * Reliability

Highlights

Highlights of GAO-12-479, a report to the Committee on Armed Services, House of Representatives

ELECTRONIC WARFARE

DOD Actions Needed to Strengthen Management and Oversight

I0471932

Why GAO Did This Study

DOD has committed billions of dollars to developing, maintaining, and employing warfighting capabilities that rely on access to the electromagnetic spectrum. According to DOD, electronic warfare capabilities play a critical and potentially growing role in ensuring the U.S. military's access to and use of the electromagnetic spectrum. GAO was asked to assess the extent to which DOD (1) developed a strategy to manage electronic warfare and (2) planned, organized, and implemented an effective governance structure to oversee its electronic warfare policy and programs and their relationship to cyberspace operations. GAO analyzed policies, plans, and studies related to electronic warfare and cyberspace operations and interviewed cognizant DOD officials.

What GAO Recommends

GAO recommends that DOD should (1) include in its future electronic warfare strategy reports to Congress certain key characteristics, including performance measures, key investments and resources, and organizational roles and responsibilities; (2) define objectives and issue an implementation plan for the Joint Electromagnetic Spectrum Control Center; and (3) update key departmental guidance to clearly define oversight roles, responsibilities, and coordination for electronic warfare management, and the relationship between electronic warfare and cyberspace operations. DOD generally concurred with these recommendations, except that the strategy should include performance measures. GAO continues to believe this recommendation has merit.

View GAO-12-479. For more information, contact Brian J. Lepore at (202) 512-4523 or leporeb@gao.gov.

What GAO Found

The Department of Defense (DOD) developed an electronic warfare strategy, but it only partially addressed key characteristics that GAO identified in prior work as desirable for a national or defense strategy. The National Defense Authorization Act for Fiscal Year 2010 requires DOD to submit to the congressional defense committees an annual report on DOD's electronic warfare strategy for each of fiscal years 2011 through 2015. DOD issued its fiscal year 2011 and 2012 strategy reports to Congress in October 2010 and November 2011, respectively. GAO found that DOD's reports addressed two key characteristics: (1) purpose, scope, and methodology and (2) problem definition and risk assessment. However, DOD only partially addressed four other key characteristics of a strategy, including (1) resources, investments, and risk management and (2) organizational roles, responsibilities, and coordination. For example, the reports identified mechanisms that could foster coordination across the department and identified some investment areas, but did not fully identify implementing parties, delineate roles and responsibilities for managing electronic warfare across the department, or link resources and investments to key activities. Such characteristics can help shape policies, programs, priorities, resource allocation, and standards in a manner that is conducive to achieving intended results and can help ensure that the department is effectively managing electronic warfare.

DOD has taken steps to address a critical electronic warfare management gap, but it has not established a departmentwide governance framework for electronic warfare. GAO previously reported that effective and efficient organizations establish objectives and outline major implementation tasks. In response to a leadership gap for electronic warfare, DOD is establishing the Joint Electromagnetic Spectrum Control Center under U.S. Strategic Command as the focal point for joint electronic warfare. However, because DOD has yet to define specific objectives for the center, outline major implementation tasks, and define metrics and timelines to measure progress, it is unclear whether or when the center will provide effective departmentwide leadership and advocacy for joint electronic warfare. In addition, key DOD directives providing some guidance for departmentwide oversight of electronic warfare have not been updated to reflect recent changes. For example, DOD's primary directive concerning electronic warfare oversight was last updated in 1994 and identifies the Under Secretary of Defense for Acquisition, Technology, and Logistics as the focal point for electronic warfare. The directive does not define the center's responsibilities in relation to the office, including those related to the development of the electronic warfare strategy and prioritizing investments. In addition, DOD's directive for information operations, which is being updated, allocates electronic warfare responsibilities based on the department's previous definition of information operations, which had included electronic warfare as a core capability. DOD's oversight of electronic warfare capabilities may be further complicated by its evolving relationship with computer network operations, which is also an information operations-related capability. Without clearly defined roles and responsibilities and updated guidance regarding oversight responsibilities, DOD does not have reasonable assurance that its management structures will provide effective departmentwide leadership for electronic warfare activities and capabilities development and ensure effective and efficient use of its resources.

_____ United States Government Accountability Office

Contents

Abbreviations

DOD	Department of Defense
JEMSCC	Joint Electromagnetic Spectrum Control Center

July 9, 2012

The Honorable Howard P. "Buck" McKeon
Chairman
The Honorable Adam Smith
Ranking Member
Committee on Armed Services
House of Representatives

The Department of Defense (DOD) is increasingly dependent on access to the electromagnetic spectrum—the full range of all possible frequencies of electromagnetic radiation, including frequency ranges such as radio, microwave, infrared, visible, ultraviolet, X-rays, and gamma rays—for a variety of military uses, such as communicating, navigating, information gathering and sensing, and targeting. DOD has committed billions of dollars developing, maintaining, and employing warfighting capabilities that rely on access to the electromagnetic spectrum—including precision-guided munitions and command, control, and communications systems. DOD ensures control of the electromagnetic spectrum through the coordinated implementation of joint electromagnetic spectrum operations, which includes electronic warfare and spectrum management activities, with other lethal and nonlethal operations that enable freedom of action in the electromagnetic operational environment. Electronic warfare, which is the use of electromagnetic energy and directed energy to control the electromagnetic spectrum or to attack the enemy, is essential for protection of friendly operations and denying adversary operations within the electromagnetic spectrum throughout the operational environment. As we previously reported, DOD's investments are projected to total more than $17.6 billion from fiscal years 2007 through 2016 for the development and procurement of new and updated fixed-wing airborne electronic attack systems alone, which are one element of electronic warfare.[1]

According to DOD, the U.S. military's access to and use of the electromagnetic spectrum is facing rapidly evolving challenges and increased vulnerabilities due to the increasing quality and availability of

---

[1] GAO, *Airborne Electronic Attack: Achieving Mission Objectives Depends on Overcoming Acquisition Challenges*, GAO-12-175 (Washington, D.C.: Mar. 29, 2012).

electronic warfare capabilities to both state and non-state actors. Also, DOD has reported that electronic warfare capabilities, which play a critical and potentially growing role as an enabler for military operations, are currently stressed and will remain so in the future. Moreover, according to DOD, near-peer competitors, primarily Russia and China, have fully recognized the critical nature of electromagnetic spectrum control in military operations.[2] There also has been recognition among near-peer competitors of the relationship between electronic warfare and cyberspace operations, which includes computer network operations.[3] For example, as noted in the U.S.-China Economic and Security Review Commission's 2009 report to Congress, China's Integrated Network Electronic Warfare concept incorporates elements of cyberspace operations in tandem with elements of traditional electronic warfare, and advocates for the employment of traditional electronic warfare operations—such as the jamming of radars and communications systems—in coordination with cyberspace attack operations.

DOD has identified persistent electronic warfare capability gaps, and these shortfalls have been consistently highlighted by the combatant commands as some of their highest warfighting priorities. According to a Center for Strategic and International Studies report, the U.S. Strategic Command identified 34 capability gaps affecting electronic warfare, including a lack of leadership across the department.[4] This lack of leadership was identified as the most critical gap. In our recent report on DOD's airborne electronic attack capabilities, we found that DOD is

---

[2] Potential near-peer adversaries include countries capable of waging large-scale conventional war on the United States. These nation-states are characterized as having nearly comparable diplomatic, informational, military, and economic capacity to the United States.

[3] DOD defines cyberspace operations, which includes computer network operations, as the employment of cyberspace capabilities where the primary purpose is to achieve military objectives or effects in or through cyberspace. DOD documents that discuss the relationship between electronic warfare and cyberspace operations use several different cyber-related terms, including cyberspace, cyber operations, computer network operations, and computer network attack. In addition, according to DOD, the definition of information operations includes the term computer network operations because it is an information operations-related capability. To provide clarity in this report, we generally use the term cyberspace operations in our discussion of the relationship between electronic warfare and cyberspace operations and computer network operations in our discussions concerning information operations-related capabilities.

[4] Center for Strategic and International Studies, *Organizing for Electro-Magnetic Spectrum Control* (Washington, D.C.: May 2010).

developing multiple systems which provide similar capabilities, and that the lack of leadership may undermine DOD's ability to consolidate these systems.[5] Specifically, we found that all four military services within the Department of Defense are separately acquiring new airborne electronic attack systems, but that opportunities may exist to consolidate some current service-specific acquisition efforts. With the prospect of slowly growing or flat defense budgets for years to come, the department must get better returns on its weapon system investments and find ways to deliver more capability to the warfighter for less than it has in the past. Therefore, we recommended that the Secretary of Defense conduct program reviews for certain new, key systems; determine the extent to which the most pressing capability gaps can be met and take steps to fill them; align service investments in science and technology with the departmentwide electronic warfare priority; and review the capabilities provided by certain existing and planned systems to ensure investments do not overlap. DOD generally concurred with our recommendations.

You requested that we examine several issues related to DOD's electronic warfare capabilities. In March 2012, we issued a report on DOD's current and planned airborne electronic attack capabilities and investment strategies.[6] In this current review, we examined DOD's approach to governing electronic warfare and the relationship between electronic warfare and cyberspace operations. Specifically, we examined the extent to which DOD has (1) developed a strategy to manage electronic warfare and (2) planned, organized, and implemented an effective governance structure to oversee its electronic warfare policy and programs, and their relationship to cyberspace operations.

To assess the extent to which DOD has developed a strategy to manage electronic warfare, we compared information found in DOD's two electronic warfare strategy reports to Congress with key characteristics of strategies identified by GAO in prior work, and interviewed relevant officials. To assess the extent to which DOD has planned, organized, and implemented an effective governance structure to oversee its electronic warfare policy and programs and their relationship to cyberspace

---

[5] GAO-12-175. For additional information, see GAO, *2012 Annual Report: Opportunities to Reduce Duplication, Overlap and Fragmentation, Achieve Savings, and Enhance Revenue*, GAO-12-342SP (Washington, D.C.: Feb. 28, 2012).

[6] GAO-12-175.

operations, we reviewed DOD directives and policies, and the roles and responsibilities of the Under Secretary of Defense for Policy; Under Secretary of Defense for Acquisition, Technology, and Logistics; and U.S. Strategic Command. Additionally, we reviewed and analyzed information found in policy documents along with information from relevant meetings with DOD officials against DOD's directives regarding electronic warfare. We also interviewed cognizant officials and reviewed DOD policies, doctrine, reports, plans, and concepts of operation, and outside studies that discuss the relationship between electronic warfare and cyberspace operations. See Appendix I for details on our scope and methodology.

We conducted this performance audit from July 2011 to July 2012 in accordance with generally accepted government auditing standards. Those standards require that we plan and perform the audit to obtain sufficient, appropriate evidence to provide a reasonable basis for our findings and conclusions based on our audit objectives. We believe that the evidence obtained provides a reasonable basis for our findings and conclusions based on our audit objectives.

Background

Control and Use of the Electromagnetic Spectrum

In modern warfare, military forces are heavily dependent upon access to the electromagnetic spectrum for successful operations. Communications with friendly forces and detection, identification, and targeting of enemy forces, among other tasks, are all reliant upon the ability to operate unhindered in the spectrum. For this reason, control of the electromagnetic spectrum is considered essential to carrying out military operations.[7] Figure 1 illustrates the electromagnetic spectrum and some examples of military uses at various frequencies. For example, infrared or thermal imaging technology senses heat emitted by a person or an object and creates an image. Sensor systems utilize this technology to provide

---

[7] According to DOD, electromagnetic spectrum control is freedom of action in the electromagnetic operational environment, which is achieved through the coordinated implementation of joint electromagnetic spectrum operations, which includes electronic warfare, with other lethal and nonlethal operations impacting the electromagnetic operational environment. See Joint Chiefs of Staff, Joint Publication 3-13.1, *Electronic Warfare* (Feb. 8, 2012).

the advantage of seeing not only at night but also through smoke, fog, and other obscured battlefield conditions.

Figure 1: Electromagnetic Spectrum and Uses

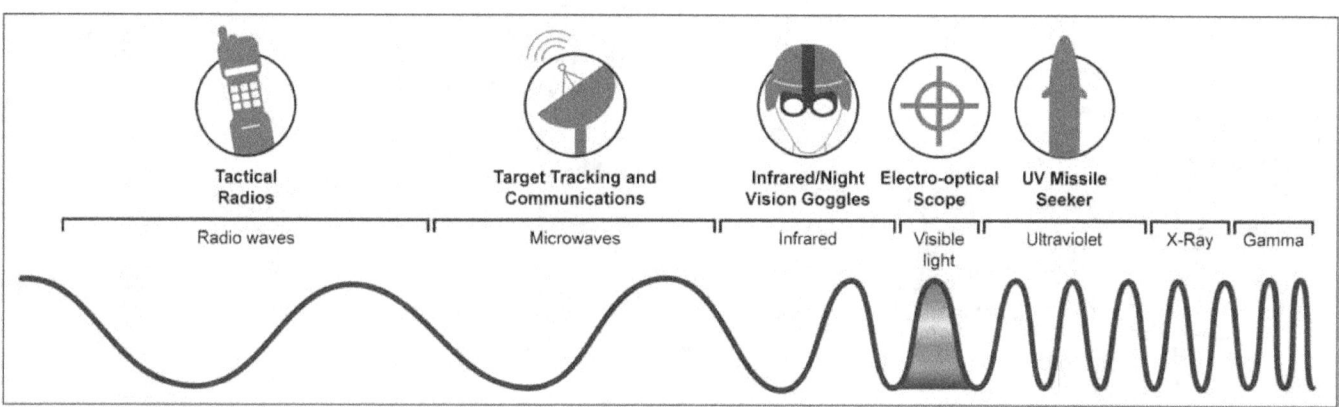

Source: GAO analysis based on DOD information.

Electronic Warfare

DOD defines electronic warfare as any military action involving the use of electromagnetic and directed energy to control the electromagnetic spectrum or to attack the enemy. The purpose of electronic warfare is to secure and maintain freedom of action in the electromagnetic spectrum for friendly forces and to deny the same for the adversary. Traditionally, electronic warfare has been composed of three primary activities:

- Electronic attack: use of electromagnetic, directed energy, or antiradiation weapons to attack personnel, facilities, or equipment with the intent of degrading, neutralizing, or destroying enemy combat capability. Electronic attack can be used offensively, such as jamming enemy communications or jamming enemy radar to suppress its air defenses, and defensively, such as deploying flares.
- Electronic protection: actions to protect personnel, facilities, and equipment from any effects of friendly, neutral, or enemy use of the electromagnetic spectrum, as well as naturally occurring phenomena that degrade, neutralize, or destroy friendly combat capability.
- Electronic warfare support: actions directed by an operational commander to search for, intercept, identify, and locate sources of radiated electromagnetic energy for the purposes of immediate threat recognition, targeting, and planning; and conduct of future operations.

GAO-12-479 Electronic Warfare

Figure 2: Examples of Electronic Warfare Capabilities

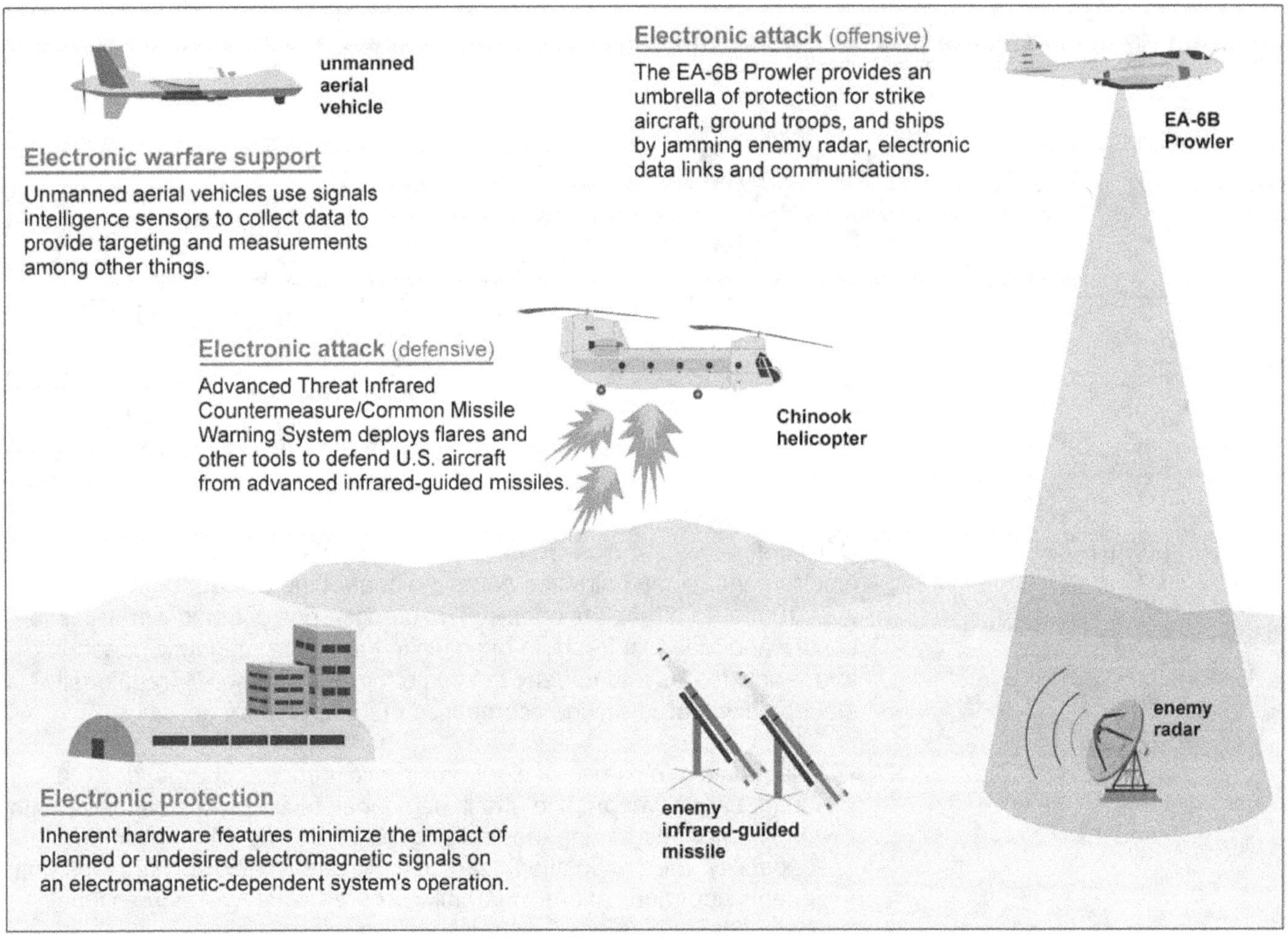

unmanned aerial vehicle

Electronic warfare support

Unmanned aerial vehicles use signals intelligence sensors to collect data to provide targeting and measurements among other things.

Electronic attack (offensive)

The EA-6B Prowler provides an umbrella of protection for strike aircraft, ground troops, and ships by jamming enemy radar, electronic data links and communications.

EA-6B Prowler

Electronic attack (defensive)

Advanced Threat Infrared Countermeasure/Common Missile Warning System deploys flares and other tools to defend U.S. aircraft from advanced infrared-guided missiles.

Chinook helicopter

Electronic protection

Inherent hardware features minimize the impact of planned or undesired electromagnetic signals on an electromagnetic-dependent system's operation.

enemy infrared-guided missile

enemy radar

Source: GAO analysis based on DOD information.

Information Operations

Electronic warfare is employed to create decisive stand-alone effects or to support military operations, such as information operations and cyberspace operations. According to DOD, information operations are the integrated employment, during military operations, of information-related capabilities in concert with other lines of operation to influence, disrupt, corrupt, or usurp the decision-making of adversaries and potential adversaries while protecting our own. Information-related capabilities can include, among others, electronic warfare, computer network operations, military deception, operations security, and military information support operations (formerly psychological operations). Electronic warfare contributes to the success of information operations by using offensive and defensive tactics and techniques in a variety of combinations to shape, disrupt, and exploit adversarial use of the electromagnetic spectrum while protecting U.S. and allied freedom of action.

Cyberspace Operations

Since cyberspace requires both wired and wireless links to transport information, both offensive and defensive cyberspace operations may require use of the electromagnetic spectrum. According to DOD, cyberspace operations are the employment of cyberspace capabilities where the primary purpose is to achieve military objectives or effects through cyberspace, which include computer network operations, among others. Computer network operations include computer network attack, computer network defense, and related computer network exploitation-enabling operations. Electronic warfare and cyberspace operations are complementary and have potentially synergistic effects. For example, an electronic warfare platform may be used to enable or deter access to a computer network.

U.S. Strategic Command Joint Electronic Warfare Activities

U.S. Strategic Command (Strategic Command) has been designated since 2008 as the advocate for joint electronic warfare. Strategic Command officials stated that, in the past, the primary office for electronic warfare expertise—the Joint Electronic Warfare Center—had several different names and was aligned under several different organizations, such as the Joint Forces Command and the U.S. Space Command.

According to Strategic Command officials, in addition to the Joint Electronic Warfare Center, the command employs electronic warfare experts in its non-kinetic operations staff and in the Joint Electromagnetic Preparedness for Advanced Combat organization. According to Strategic Command officials, the Joint Electronic Warfare Center is the largest of the three organizations and employs approximately 60 military and civilian electronic warfare personnel and between 15 and 20 contractors. Strategic Command officials stated that the Joint Electronic Warfare Center was created as a DOD center of excellence for electronic warfare and has electronic warfare subject matter experts. The center provides planning and technical support not only to Strategic Command but to other combatant commands and organizations, such as U.S. Central Command, U.S. European Command, U.S. Pacific Command, and the Department of Homeland Security. The Joint Electronic Warfare Center also provides assistance with requirements generation to the military services.

DOD Developed an Electronic Warfare Strategy, but Only Partially Addressed Key Desirable Strategy Characteristics

DOD developed an electronic warfare strategy, but only partially addressed key strategy characteristics identified as desirable in prior work by GAO. The *National Defense Authorization Act for Fiscal Year 2010* requires the Secretary of Defense to submit to the congressional defense committees an annual report on DOD's electronic warfare strategy for each of fiscal years 2011 through 2015.[8] Each annual report is to be submitted at the same time the President submits the budget to Congress and is to contain, among other things, a description and overview of DOD's electronic warfare strategy and the organizational structure assigned to oversee the development of the department's electronic warfare strategy, requirements, capabilities, programs, and projects.[9] In response to this legislative requirement, the Office of the Under Secretary of Defense for Acquisition, Technology, and Logistics issued DOD's 2011 and 2012 fiscal year strategy reports to Congress in October 2010 and November 2011, respectively.[10]

We previously reported that it is desirable for strategies to delineate six key characteristics, including organizational roles and responsibilities for implementing parties as well as performance measures to gauge

---

[8] Pub. L. No. 111-84, § 1053 (a).

[9] Pub. L. No. 111-84, § 1053 (b) delineates several other requirements for the strategy, including a list of electronic warfare acquisition programs and research and development projects with associated program or project information.

[10] Office of the Under Secretary of Defense for Acquisition, Technology, and Logistics, Report to the Congressional Defense Committees. *Electronic Warfare Strategy of the Department of Defense* (Washington, D.C.: October 2010). The fiscal year 2012 report is classified. As of July 2012, the fiscal year 2013 electronic warfare strategy report was still being drafted, according to officials from the Office of the Under Secretary of Defense for Acquisition, Technology, and Logistics.

results.[11] The key characteristics of an effective strategy can aid responsible parties in further developing and implementing the strategy, enhance the strategy's usefulness in resource and policy decisions, and better ensure accountability. The six characteristics are: (1) purpose, scope, and methodology; (2) problem definition and risk assessment; (3) goals, subordinate objectives, activities, and performance measures; (4) resources, investments, and risk management; (5) organizational roles, responsibilities, and coordination; and (6) integration and implementation.

As illustrated in Figure 3, we found that DOD's reports addressed two key characteristics, but only partially addressed four other key characteristics of a strategy. For example, the strategy reports to Congress included elements of characteristics, such as a goal and objectives, but did not fully identify implementing parties, delineate roles and responsibilities for managing electronic warfare across the department, or identify outcome-related performance measures that could guide the implementation of electronic warfare efforts and help ensure accountability. Similarly, the reports provided acquisition program and research and development project data, but did not target resources and investments at some key activities associated with implementing the strategy. When investments are not tied to strategic goals and priorities, resources may not be used effectively and efficiently. Our past work has shown that such characteristics can help shape policies, programs, priorities, resource

---

[11] See GAO, *Combating Terrorism: Evaluation of Selected Characteristics in National Strategies Related to Terrorism*, GAO-04-408T (Washington, D.C.: Feb. 3, 2004). This testimony identified six characteristics of an effective strategy. While these characteristics were identified in our past work as desirable components of national-level strategies, we determined that they also are relevant to strategies of varying scopes, including defense strategies involving complex issues. For example, identifying organizational roles, responsibilities and coordination mechanisms is key to allocating authority and responsibility for implementing a strategy. Further, goals, objectives, and performance measures provide concrete guidance for implementing a strategy, allowing implementing parties to establish priorities and milestones, and providing them with the flexibility necessary to pursue and achieve those results within a reasonable timeframe. Full descriptions of these characteristics are contained in appendix II. See also GAO, *Influenza Pandemic: DOD Has Taken Important Actions to Prepare, but Accountability, Funding, and Communications Need to be Clearer and Focused Departmentwide*, GAO-06-1042 (Washington, D.C.: Sept. 21, 2006) ; GAO, *Defense Space Activities: National Security Space Strategy Needed to Guide Future DOD Space Efforts*, GAO-08-431R (Washington, D.C.: Mar. 27, 2008); and GAO, *Intelligence, Surveillance, and Reconnaissance: DOD Needs a Strategic, Risk-Based Approach to Enhance Its Maritime Domain Awareness*, GAO-11-621 (Washington, D.C.: June 20, 2011).

allocations, and standards in a manner that is conducive to achieving intended results.[12]

Figure 3: Extent to Which DOD's Fiscal Year 2011 Electronic Warfare Strategy Report Addressed Key Desirable Strategy Characteristics Identified by GAO

Desirable Characteristics and Summary Analyses	GAO assessment
Purpose, scope, and methodology	
Addressed	●
Problem definition and risk assessment	
Addressed	●
Goals, subordinate objectives, activities, and performance measures	
Partially addressed—The report identifies the goal, objectives, and several activities associated with the strategy, but does not identify outcome-related performance measures that could help ensure accountability.	◒
Resources, investments, and risk management	
Partially addressed—The report refers to investment areas and provides data on acquisition programs and research and development projects, but does not target resources and investments at key activities associated with implementing the strategy, or balance risk against cost.	◒
Organizational roles, responsibilities, and coordination	
Partially addressed—The report provides an overview of some electronic warfare activities and identifies mechanisms that could be used to foster coordination across the department, but it does not fully identify which entities will be implementing the strategy or discuss implementing entities' roles and responsibilities.	◒
Integration and implementation	
Partially addressed—The report describes how the strategy supports the objectives and activities of the National Defense Strategy, but does not address implementation plans for the strategy.	◒

● Addressed
◒ Partially addressed
○ Does not address

Source: GAO analysis of DOD's fiscal year 2011 *Electronic Warfare Strategy of the Department of Defense* report to Congress.

Notes: Summary analysis information is provided only in cases where we have determined the strategy partially addressed a characteristic. The strategy "addressed" a characteristic when the strategy explicitly cited all elements of a characteristic, even if it lacked specificity and details and thus could be improved upon. The strategy "partially addressed" a characteristic when the strategy explicitly cited some, but not all, elements of a characteristic. Within our designation of "partially addressed," there may be wide variation between a characteristic for which most of the elements were addressed and a characteristic for which few of the elements of the characteristic were addressed. The strategy "did not address" a characteristic when the strategy did not explicitly cite or discuss any elements of a characteristic, and/or any implicit references were either too vague or general.

---

[12] GAO-04-408T.

DOD's fiscal year 2011 report is described here because the fiscal year 2012 report, issued in November 2011, is classified. However, unclassified portions of this document note that the fiscal year 2011 report remains valid as the base DOD strategy and that the fiscal year 2012 report updates its predecessor primarily to identify ongoing efforts to improve DOD's electronic warfare capabilities and to provide greater specificity to current threats. The fiscal year 2011 *Electronic Warfare Strategy of the Department of Defense* report (electronic warfare strategy report)—the base electronic warfare strategy—addressed two and partially addressed four of six desirable characteristics of a strategy identified by GAO. There may be considerable variation in the extent to which the strategy addressed specific elements of those characteristics that were determined by GAO to be partially addressed. Our analysis of the fiscal year 2011 report's characteristics is as follows.

- **Purpose, scope and methodology: *Addressed.*** The fiscal year 2011 electronic warfare strategy report identifies the purpose of the strategy, citing as its impetus section 1053 of the *National Defense Authorization Act for Fiscal Year 2010*, and articulates a maturing, twofold strategy focused on integrating electronic warfare capabilities into all phases and at all levels of military operations, as well as developing, maintaining, and protecting the maneuver space within the electromagnetic spectrum necessary to enable military capabilities. The report's scope also encompasses data on acquisition programs and research and development projects. Additionally, the report includes some methodological information by citing a principle that guided its development. Specifically the report states that a key aspect of the strategy is the concept of the electromagnetic spectrum as maneuver space.

- **Problem definition and risk assessment: *Addressed.*** The fiscal year 2011 electronic warfare strategy report defines the problem the strategy intends to address, citing the challenges posed to U.S. forces by potential adversaries' increasingly sophisticated technologies, the military's increased dependence on the electromagnetic spectrum, and the urgent need to retain and expand remaining U.S. advantages. The report also assesses risk by identifying threats to, and vulnerabilities of critical operations, such as Airborne Electronic Attack and self-protection countermeasures.

- **Goals, subordinate objectives, activities, and performance measures: *Partially Addressed.*** The fiscal year 2011 electronic warfare strategy report communicates an overarching goal of enabling electromagnetic spectrum maneuverability and cites specific objectives, such as selectively denying an adversary's use of the spectrum and preserving U.S. and allied forces' ability to maneuver within the spectrum. The report also identifies key activities associated with the strategy, including developing (1) coherent electronic warfare organizational structures and leadership, (2) an enduring and sustainable approach to continuing education, and (3) capabilities to implement into electronic warfare systems. The report does not identify performance measures that could be used to gauge results and help ensure accountability.

- **Resources, investments, and risk management:** *Partially Addressed.* The fiscal year 2011 electronic warfare strategy report broadly targets resources and investments by emphasizing the importance of continued investment in electronic attack, electronic protection, and electronic support capabilities. The report also notes some of the associated risks in these areas, calling for new methods of ensuring U.S. control over the electromagnetic spectrum in light of the adversary's advances in weapons and the decreasing effectiveness of traditional lines of defense, such as airborne electronic attack and self-protection countermeasures. The report identifies some of the costs associated with the strategy by providing acquisition program and research and development project and cost data, and notes that part of the strategy is to identify and track investments in electronic warfare systems, which often are obscured within the development of the larger weapons platforms they typically support. However, the strategy does not target investments by balancing risk against costs, or discuss other costs associated with implementing the strategy by, for example, targeting resources and investments at key activities, such as developing electronic warfare organizational structures and leadership and developing an enduring and sustainable approach to continuing education.

- **Organizational roles, responsibilities, and coordination:** *Partially Addressed.* The fiscal year 2011 electronic warfare strategy report provides an overview of past and ongoing electronic warfare activities within the military services and DOD, and identifies several mechanisms that have or could be used to foster coordination across the department. For example, it outlines the Army's efforts to create a new career field for electronic warfare officers and the Office of the Under Secretary of Defense for Acquisition, Technology and Logistics' electronic warfare integrated planning team.[13] However, the report does not fully identify the departmental entities responsible for implementing the strategy, discuss the roles and responsibilities of implementing parties, or specify implementing entities' relationships in terms of leading, supporting, and partnering.[14]

- **Integration and implementation:** *Partially Addressed.* The fiscal year 2011 electronic warfare strategy report describes the department's approach to ensuring maneuverability within the electromagnetic spectrum, thus supporting *National Defense Strategy* objectives that rely on use and control of the spectrum. The strategy's overarching aim of ensuring electromagnetic spectrum maneuverability also is consistent with concepts contained in the department's electromagnetic spectrum strategy documents—which collectively emphasize the importance of assured spectrum access.[15] The strategy does not, however, discuss the department's plans for implementing the strategy.

DOD's electronic warfare strategy reports were issued in response to the *National Defense Authorization Act for Fiscal Year 2010* and were not specifically required to address all the characteristics we consider to be desirable for an effective strategy. Additionally, DOD's fiscal year 2011 report states that the strategy is still maturing and that subsequent reports to Congress will refine the department's vision. Nonetheless, we consider it useful for DOD's electronic warfare strategy to address each of the characteristics we have identified in order to provide guidance to the entities responsible for implementing DOD's strategy and to enhance the strategy's utility in resource and policy decisions—particularly in light of the diffuse nature of DOD's electronic warfare programs and activities, as well as the range of emerging technical, conceptual, and organizational challenges and changes in this area. Further, in the absence of clearly

---

[13] According to DOD officials, the electronic warfare integrated planning team was established by the Office of the Under Secretary of Defense for Acquisition, Technology and Logistics to develop guidelines for electronic warfare investment strategy.

[14] By "partnering," we refer to shared, or joint, responsibilities among implementing parties where there is otherwise no clear or established hierarchy of lead and support functions.

[15] See Department of Defense, *Strategic Spectrum Plan* (February 2008); and Assistant Secretary of Defense (Networks and Information Integration) Department of Defense Chief Information Officer, *Department of Defense Net-Centric Spectrum Management Strategy* (Aug. 3, 2006).

defined roles and responsibilities, and other elements of key characteristics, such as measures of performance in meeting goals and objectives, entities responsible for implementing DOD's strategy may lack the guidance necessary to establish priorities and milestones, thereby impeding their ability to achieve intended results within a reasonable time frame. As a result, DOD lacks assurance that its electronic warfare programs and activities are aligned with strategic priorities and are managed effectively. For example, without an effective strategy, DOD is limited in its ability to reduce the potential for unnecessary overlap in the airborne electronic attack acquisition activities on which we have previously reported.

DOD Has Not Established an Effective Departmentwide Governance Framework for Managing and Overseeing Electronic Warfare

DOD has taken some steps to address a critical leadership gap identified in 2009, but it has not established a departmentwide governance framework for planning, directing, and controlling electronic warfare activities. DOD is establishing a Joint Electromagnetic Spectrum Control Center (JEMSCC) under Strategic Command in response to the leadership gap for electronic warfare. However, DOD has not documented the objectives or implementation tasks and timeline for the JEMSCC. In addition, DOD has not updated key guidance to reflect recent policy changes regarding electronic warfare management and oversight roles and responsibilities. For example, it is unclear what the JEMSCC's role is in relation to other DOD organizations involved in the management of electronic warfare, such as the Office of the Under Secretary of Defense for Acquisition, Technology, and Logistics. Moreover, we found that DOD may face challenges in its oversight of electronic warfare as a result of the evolving relationship between electronic warfare and cyberspace operations.

DOD Actions Have Not Fully Addressed a Critical Leadership Gap

DOD has taken some steps to address a critical leadership gap by establishing the JEMSCC under Strategic Command. However, because DOD has yet to define specific objectives for the center, outline major implementation tasks, and define metrics and timelines to measure progress, it is unclear to what extent the center will address the identified existing leadership deficiencies. The Center for Strategic and International Studies reported insufficient leadership as the most critical among 34 capability gaps affecting electronic warfare. As a result of the absence of leadership, the department was significantly impeded from both identifying departmentwide needs and solutions and eliminating potentially unnecessary overlap among the military services' electronic warfare acquisitions. Specifically, the department lacked a joint leader and advocate with the authority to integrate and influence electronic warfare capabilities development, to coordinate internal activities, and to represent those activities and interests to outside organizations. Mitigating the leadership gap was identified not only as the highest priority, but also a prerequisite to addressing the other 33 gaps.

The Center for Strategic and International Studies report was one of two parallel studies commissioned by the Joint Requirements Oversight Council[16] to assess potential organizational and management solutions to the leadership gap.[17] These studies considered a number of options, including an organization under the Deputy Secretary of Defense, an activity controlled by the Chairman of the Joint Chiefs of Staff, and an organization at Strategic Command. As a result of these studies, in January 2011, DOD initiated efforts to establish the JEMSCC under Strategic Command as the focal point of joint electronic warfare advocacy. This solution was chosen, in part, in recognition of Strategic Command's resident electronic warfare expertise as well as its already assigned role as an electronic warfare advocate.[18]

In January 2011, the Joint Requirements Oversight Council directed Strategic Command to develop an implementation plan for the electronic

---

[16] The Joint Requirements Oversight Council, among other things, assists the Chairman of the Joint Chiefs of Staff in (1) identifying, assessing, establishing priority levels for, and validating joint military requirements, including existing systems and equipment, to meet the *National Military Strategy*; (2) considering trade-offs among cost, schedule, and performance objectives for joint military requirements; and (3) reviewing the estimated level of resources required to fulfill each joint military requirement, and establishing an objective for the overall period of time within which an initial operational capability should be delivered to meet each joint military requirement.

[17] Specifically, in October 2009, the Joint Requirements Oversight Council tasked Strategic Command and the now disestablished U.S. Joint Forces Command to assess both the technical issues related to the electronic warfare problem and the organizational structure and management approach required to respond to emerging electromagnetic spectrum threats. In response, Strategic Command and U.S. Joint Forces Command produced a classified report providing potential organizational solutions. As part of this process, the Center for Strategic and International Studies conducted its review as an independent analysis that also provided organizational alternatives. See Center for Strategic and International Studies, *Organizing for Electro-Magnetic Spectrum Control* (Washington, D.C.: May 2010).

[18] See DOD, *Unified Command Plan* (Washington, D.C.: Dec. 17, 2008), which establishes the missions, responsibilities, and geographic areas of responsibility among the combatant commanders. This plan also tasks the Commander of Strategic Command with advocating for electronic warfare capabilities.

warfare center to be submitted for council approval no later than May 2011. The plan was to delineate (1) the center's mission, roles, and responsibilities; (2) command and control, reporting, and support relationships with combatant commands, military services, and U.S. Government departments and agencies; and (3) minimum requirements to achieve initial operational capability and full operational capability. The Joint Requirements Oversight Council subsequently approved an extension of the center's implementation plan submission to August 2011. Subsequently, in December 2011, the oversight council issued a memorandum that closed the requirement to submit an implementation plan to the council and stated that Strategic Command had conducted an internal reorganization and developed a center to perform the functions identified in the internal DOD study.

In December 2011, Strategic Command issued an operations order that defined the JEMSCC as the primary focal point for electronic warfare, supporting DOD advocacy for joint electronic warfare capability requirements, resources, strategy, doctrine, planning, training, and operational support. This order provided 22 activities that the center is to perform. Federal internal control standards require that organizations establish objectives and clearly define key areas of authority and responsibility.[19] In addition, best practices for strategic planning have shown that effective and efficient operations require detailed plans outlining major implementation tasks and defined metrics and timelines to measure progress.[20] Moreover, the independent study prepared for DOD similarly emphasized the importance of clearly defining the center's authorities and responsibilities, noting that the center's success would hinge, in part, on specifying how it is expected to relate to the department as a whole as well as its expected organizational outcomes. However, as of March 2012, Strategic Command had not issued an implementation plan or other documentation that defines the center's objectives and outlines major implementation tasks, metrics, and timelines to measure progress. Strategic Command officials told us in February 2012 that an implementation plan had been drafted, but that there were no timelines

---

[19] GAO, *Standards for Internal Control in the Federal Government*, GAO/AIMD-00-21.3.1 (Washington, D.C.: Nov. 1999).

[20] For example, see GAO, *Reserve Forces: Army Needs to Finalize an Implementation Plan and Funding Strategy for Sustaining an Operational Reserve Force*, GAO-09-898 (Washington, D.C.: Sept. 17, 2009).

for the completion of the implementation plan or a projection for when the center would reach its full operational capability. As a result, it remains unclear whether or when the JEMSCC will provide effective departmentwide leadership and advocacy for electronic warfare, and influence resource decisions related to capability development.

According to officials from Strategic Command, the JEMSCC will consist of staff from Strategic Command's Joint Electronic Warfare Center at Lackland Air Force Base, Texas, and the Joint Electromagnetic Preparedness for Advanced Combat organization, at Nellis Air Force Base, Nevada.[21] These officials stated that while each of JEMSCC's component groups' missions will likely evolve as the center matures, the JEMSCC components would continue prior support activities, such as the Joint Electronic Warfare Center's support to other combatant commands through its Electronic Warfare Planning and Coordination Cell—a rapid deployment team that provides electronic warfare expertise and support to build electronic warfare capacity. Figure 4 depicts the JEMSCC's organizational construct.

---

[21] The Joint Electronic Warfare Center, prior to incorporation in the JEMSCC, was tasked with integrating current and emerging joint electronic warfare effects to ensure spectrum control for global military operations. The center provided near-term operational solutions and advocated for long-term electromagnetic capabilities. The Joint Electromagnetic Preparedness for Advanced Combat organization was charged with advancing and improving joint warfighter effectiveness and combat capability by conducting vulnerability assessments of electromagnetic spectrum dependent capabilities, architectures, technologies, and tactics, techniques, and procedures.

GAO-12-479 Electronic Warfare

Figure 4: Joint Electromagnetic Spectrum Control Center under Strategic Command

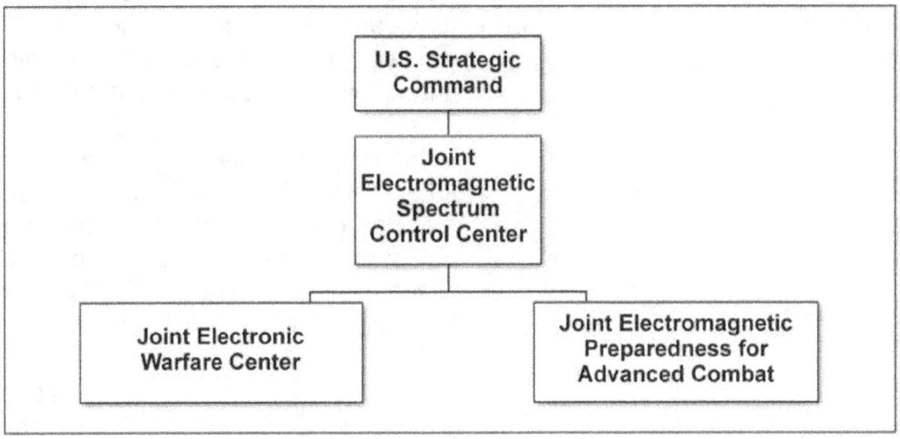

```
                    ┌──────────────────┐
                    │  U.S. Strategic  │
                    │     Command      │
                    └──────────────────┘
                             │
                    ┌──────────────────┐
                    │      Joint       │
                    │ Electromagnetic  │
                    │    Spectrum      │
                    │ Control Center   │
                    └──────────────────┘
                             │
              ┌──────────────┴──────────────┐
    ┌──────────────────┐         ┌──────────────────────┐
    │ Joint Electronic │         │ Joint Electromagnetic │
    │  Warfare Center  │         │   Preparedness for    │
    │                  │         │   Advanced Combat     │
    └──────────────────┘         └──────────────────────┘
```

Source: GAO analysis of DOD information.

DOD has yet to define objectives and issue an implementation plan for the JEMSCC; however, officials from Strategic Command stated that they anticipated continuity between the command's previous role as an electronic warfare advocate and its new leadership role, noting that advocacy was, and remains, necessary because electronic warfare capabilities are sometimes undervalued in comparison to other, kinetic capabilities.[22] For example, the JEMSCC will likely build off Strategic Command's previously assigned advocacy role, in part, by continuing to advocate for electronic warfare via the Joint Capabilities Integration and Development System process—DOD's process for identifying and developing capabilities needed by combatant commanders—and by

---

[22] Kinetic capabilities focus on destroying forces through the application of physical effects.

providing electronic warfare expertise.[23] Specifically, Strategic Command officials stated that the JEMSCC, through Strategic Command, would likely provide input to the development of joint electronic warfare requirements during the joint capabilities development process. However, combatant commands, such as Strategic Command, provide one of many inputs to this process. Further, as we have previously reported, council decisions, while influential, are advisory to acquisition and budget processes driven by military service investment priorities.[24] As a result, the JEMSCC's ability to affect resource decisions via this process is likely to be limited.

Officials we spoke with across DOD, including those from the military services and Strategic Command, recognized this challenge. Specifically, Strategic Command officials told us that for JEMSCC to influence service-level resource decisions and advocate effectively for joint electronic warfare capabilities, the JEMSCC would need to not only participate in the joint capabilities development process, but would also need authorities beyond those provided by the *Unified Command Plan*, such as the authority to negotiate with the military services regarding resource decisions. Similarly, we found that while the officials we spoke with from several DOD offices that manage electronic warfare, including offices within the military services, were unaware of the center's operational status and unclear regarding its mission, roles, and responsibilities, many also thought it to be unlikely that the JEMSCC—as a subordinate center of Strategic Command—would possess the requisite authority to advocate effectively for electronic warfare resource decisions. These

---

[23] In 2003, DOD created the Joint Capabilities Integration and Development System to guide the development of capabilities from a joint perspective. This system was established to provide the department with an integrated, collaborative process to identify and guide development of new capabilities that address the current and emerging security environment. DOD's Joint Requirements Oversight Council oversees the joint capabilities development process and participates in the development of joint requirements, which includes the identification and analysis and synthesis of capability gaps and the council's subsequent validation of capability needs. The council makes recommendations to the Chairman of the Joint Chiefs of Staff, who advises the Secretary of Defense about what capabilities to invest in as part of DOD's budget process. Before making investment decisions, the military services consider the validated capabilities during their planning, programming, and budgeting processes and make decisions among competing investments.

[24] See GAO, *Defense Management: Perspectives on the Involvement of the Combatant Commands in the Development of Joint Requirements*, GAO-11-527R (Washington, D.C.: May 20, 2011).

concerns were echoed by the independent study, which noted that the center would require strong authorities to substantially influence the allocation of other DOD elements' resources.[25]

Additionally, limited visibility across the department's electronic warfare programs and activities may impede the center's ability to advocate for electronic warfare capabilities development. Specifically, Strategic Command officials told us that they do not have access to information regarding all of the military services' electronic warfare programs and activities, particularly those that are highly classified or otherwise have special access restrictions. In addition, Strategic Command officials told us that they do not have visibility over or participate in rapid acquisitions conducted through the joint capabilities development process. In our March 2012 report on DOD's airborne electronic attack strategy and acquisitions, we reported that certain airborne electronic attack systems in development may offer capabilities that unnecessarily overlap with one another—a condition that appears most prevalent with irregular warfare systems that the services are acquiring under DOD's rapid acquisitions process.[26] The JEMSCC's exclusion from this process is likely to limit its ability to develop the departmentwide perspective necessary for effective advocacy. Moreover, in the absence of clearly defined objectives and an implementation plan outlining major implementation tasks and timelines to measure progress, these potential challenges reduce DOD's level of assurance that the JEMSCC will provide effective departmentwide leadership for electronic warfare capabilities development.

DOD Policy Documents Have Not Been Updated to Include All Oversight Roles and Responsibilities for Electronic Warfare

DOD issued two primary directives that provide some guidance for departmentwide oversight of electronic warfare. However, neither of these two directives has been updated to reflect changes in DOD's leadership structures that manage electronic warfare. Federal internal control standards require that organizations establish objectives, clearly define key areas of authority and responsibility, and establish appropriate lines of reporting to aid in the effective and efficient use of resources.[27] Additionally, those standards state that management must continually

---

[25] Center for Strategic and International Studies, *Organizing for Electro-Magnetic Spectrum Control* (Washington, D.C.: May 2010).

[26] GAO-12-175 and GAO-12-342SP.

[27] GAO/AIMD-00-21.3.1.

assess and evaluate its internal control to assure that the actions in place are effective and updated when necessary.

DOD's two primary directives that provide some guidance for departmentwide oversight of electronic warfare are:

- DOD Directive 3222.4 (*Electronic Warfare and Command and Control Warfare Countermeasures*)—Designates the Under Secretary of Defense for Acquisition (now Acquisition, Technology, and Logistics) as the focal point for electronic warfare within the department. However, the directive was issued in 1992 and updated in 1994, and does not reflect subsequent changes in policy or organizational structures. For example, the directive does not reflect the establishment of the JEMSCC under Strategic Command.

- DOD Directive 3600.01 (*Information Operations*)—Issued in 2006 and revised in May 2011, this directive provides the department with a framework for oversight of information operations, which was defined as the integrated employment of the core capabilities of electronic warfare, computer network operations, military information support operations (formerly referred to as psychological operations), military deception, and operations to influence, disrupt, corrupt, or usurp adversarial human and automated decision making while protecting that of the United States. However, the definition of oversight responsibilities for information operations has changed, and these changes have not yet been reflected in DOD Directive 3600.01.[28]

DOD Directive 3222.4 has not been updated to reflect the responsibilities for electronic warfare assigned to Strategic Command. Both the December 2008 and April 2011 versions of the *Unified Command Plan* assigned Strategic Command responsibility for advocating for joint electronic warfare capabilities.[29] Similarly, the directive has not been updated to reflect the establishment of the JEMSCC and its associated electronic warfare responsibilities. Specifically, the directive does not acknowledge that JEMSCC has been tasked by Strategic Command as the primary focal point for electronic warfare; rather, the directive

---

[28] Secretary of Defense, *Memorandum: Strategic Communication and Information Operations in the DOD* (Washington, D.C.: Jan. 25, 2011).

[29] Department of Defense, *Unified Command Plan* (Washington, D.C.: Dec. 17, 2008) and Department of Defense, *Unified Command Plan* (Washington, D.C.: Apr. 6, 2011).

designates the Under Secretary of Defense for Acquisition, Technology, and Logistics as the focal point for electronic warfare within DOD. As a result, it is unclear what JEMSCC's roles and responsibilities are in relation to those of the Under Secretary of Defense for Acquisition, Technology, and Logistics. For example, it's unclear what JEMSCC's role will be regarding development of future iterations of the DOD's electronic warfare strategy report to Congress, which is currently produced by the Office of the Under Secretary of Defense for Acquisition, Technology, and Logistics. Also it is unclear what role, if any, the JEMSCC will have in prioritizing electronic warfare investments. Moreover, the directive has not been updated to reflect the Secretary of Defense's memorandum issued in January 2011, which assigned individual capability responsibility for electronic warfare and computer network operations to Strategic Command.

DOD Directive 3600.01 provides both the Under Secretary of Defense for Acquisition, Technology, and Logistics and the Under Secretary of Defense for Intelligence with responsibilities that aid in the oversight of electronic warfare within DOD. However, pursuant to the Defense Secretary's January 2011 memo, the directive is under revision to accommodate changes in roles and responsibilities. Under the current version of DOD Directive 3600.01, the Under Secretary of Defense for Intelligence is charged with the role of Principal Staff Advisor to the Secretary of Defense for information operations. The Principal Staff Advisor is responsible for, among other things, the development and oversight of information operations policy and integration activities as well as the coordination, oversight, and assessment of the efforts of DOD components to plan, program, develop, and execute capabilities in support of information operations requirements.[30] Additionally, the current Directive 3600.01 identifies the Under Secretary of Defense for Acquisition, Technology, and Logistics as responsible for establishing specific policies for the development of electronic warfare as a core capability of information operations.

Under the requirements of DOD acquisition policy, the Under Secretary of Defense for Acquisition, Technology, and Logistics regularly collects cost,

---

[30] DOD Directive 3600.01 refers to the Principal Staff Assistant position while the Secretary's January 2011 memorandum refers to the Principal Staff Advisor position. According to DOD officials, these terms refer to the same position. We use the term Principal Staff Advisor in this report.

schedule, and performance data for major programs.[31] In some cases, the cost information of electronic warfare systems are reported as distinct programs, while in other cases, some electronic warfare systems are subcomponents of larger programs, and cost information is not regularly collected for these separate subsystems. Additionally, the Under Secretary—in coordination with the Army, the Navy, and the Air Force—is developing an implementation road map for electronic warfare science and technology. The road map is supposed to coordinate investments across DOD to accelerate the development and delivery of capabilities. The road map is expected to be completed in late summer of 2012.

The Secretary of Defense issued a memorandum in January 2011 that prompted DOD officials to begin revising DOD Directive 3600.01. The memorandum redefined information operations as "the integrated employment, during military operations, of information-related capabilities in concert with other lines of operation to influence, disrupt, corrupt, or usurp the decision-making of adversaries and potential adversaries while protecting our own." Previously, DOD defined information operations as the "integrated employment of the core capabilities of electronic warfare, computer network operations, psychological operations, military deception, and operations security, in concert with specified supporting and related capabilities, to influence, disrupt, corrupt, or usurp adversarial human and automated decision making while protecting our own." According to DOD officials, the revised definition removed the term core capabilities because it put too much emphasis on the individual core capabilities and too little emphasis on the integration of these capabilities.

Additionally, the memorandum noted that the Under Secretary of Defense for Policy began serving as the Principal Staff Advisor for information operations as of October 1, 2010, and charged the Under Secretary of Defense for Policy with revising DOD Directive 3600.01 to reflect these responsibilities. According to the memorandum, the Principal Staff Advisor is to serve as the single point of fiscal and program accountability for information operations. However, according to DOD officials, this accountability oversight covers only the integration of information operations-related capabilities and does not cover the formerly defined core capabilities of information operations, including electronic warfare

---

[31] See DOD Directive 5000.01 *The Defense Acquisition System* (Washington, D.C.: certified current as of Nov. 20, 2007) and DOD Instruction 5000.02 *Operation of the Defense Acquisition System* (Washington, D.C.: Dec. 8, 2008).

and computer network operations. For example, DOD officials stated that the Principal Staff Advisor for information operations would maintain program accountability where information operations-related capabilities were integrated but would not maintain program accountability for all information-related capabilities. However, the memorandum does not clearly describe the specific responsibilities of the Principal Staff Advisor for information operations.

The Secretary's memorandum directed the Under Secretary of Defense for Policy, together with the Undersecretary of Defense (Comptroller) and Director of Cost Analysis and Program Evaluation, to continue to work to develop standardized budget methodologies for information operations-related capabilities and activities. However, these budget methodologies would capture only data related to information operations. For example, according to Under Secretary of Defense for Policy officials, they do not collect or review electronic warfare financial data, but may review this data in the future to determine if it relates to integrated information operations efforts. Officials from the Office of the Under Secretary of Defense for Policy stated that DOD Directive 3600.01 was under revision to reflect these and other changes as directed by the Secretary's memorandum. Until the underlying directive is revised, there may be uncertainty regarding which office has the authority to manage and oversee which programs. Moreover, until this directive is updated, it is not clear where the boundaries are for oversight of electronic warfare between the Under Secretary of Defense for Policy and the Under Secretary of Defense for Acquisition, Technology, and Logistics.

Table 1 compares the oversight roles and responsibilities for electronic warfare as described in the two DOD directives and the Secretary's 2011 policy memorandum.

Table 1: Department of Defense Electronic Warfare Responsibilities

Organizations assigned electronic warfare responsibilities[a]	DOD Directive 3222.4, *Electronic Warfare*, July 31, 1992 (Incorporating Change 2, January 28, 1994)	DOD Directive 3600.01, *Information Operations*, August 14, 2006 (Incorporating Change 1, May 23, 2011) (Electronic warfare defined as one of the core capabilities of information operations)	Secretary of Defense, *Memorandum: Strategic Communication and Information Operations in the DOD*, January 25, 2011 (Electronic warfare defined as an information operations-related capability)
Office of the Under Secretary of Defense for Policy	• No responsibility assigned.	• Provide oversight of information operations planning, execution, and related policy guidance, including the establishment of an Office of the Secretary of Defense review process to assess information operations plans and programs submitted by combatant commanders.	• Assigned the Principal Staff Advisor function and responsibility for information operations oversight and management. • Tasked to revise DOD Directive 3600.01 and DOD Directive 5111.1, and other relevant policy and doctrine documents to reflect a new definition of information operations. • Assigned as the single point of fiscal and program accountability for information operations.
Office of the Under Secretary of Defense for Acquisition, Technology, and Logistics	• Focal point for electronic warfare within DOD. • Provide guidance on electronic warfare policy. • Provide oversight for development and acquisition of tactical land, sea, air, space, or undersea electronic warfare systems. • Review electronic warfare programs for duplication and maximum multi-service applications. • Provide matrix electronic warfare technical and/or management support within the Office of the Secretary of Defense on request. • Ensure that adequate science and technology programs exist for development and acquisition of electronic warfare systems.	• Establish specific policies for the development and integration of electronic warfare. • Develop and maintain a technology investment strategy to support the development, acquisition, and integration of electronic warfare capabilities. • Invest in and develop the science and technologies needed to support information operations capabilities.	• No responsibility assigned..

Organizations assigned electronic warfare responsibilities[a]	DOD Directive 3222.4, *Electronic Warfare*, July 31, 1992 (Incorporating Change 2, January 28, 1994)	DOD Directive 3600.01, *Information Operations*, August 14, 2006 (Incorporating Change 1, May 23, 2011) (Electronic warfare defined as one of the core capabilities of information operations)	Secretary of Defense, *Memorandum: Strategic Communication and Information Operations in the DOD*, January 25, 2011 (Electronic warfare defined as an information operations-related capability)
Office of the Under Secretary of Defense for Intelligence	• No responsibility assigned.	• Serve as the Principal Staff Advisor for information operations. • Develop and oversee DOD information operations policy and integration activities. • Coordinate, oversee, and assess the efforts of the DOD components to plan, program, develop, and execute capabilities in support of information operations requirements. • Establish specific policies for the development and integration of computer network operations, military deception and operational security.	• Relieved of role as the Principal Staff Advisor function and responsibility for information operations oversight and management.
Strategic Command[b]	NA[c]	• Shall integrate and coordinate DOD information operations core capabilities (including electronic warfare and computer network operations) that cross geographic areas of responsibility or across the core information operations areas.	• Assigned individual capability responsibilities for electronic warfare and computer network operations.

Source: GAO analysis of DOD documents.

Note: This table presents only those responsibilities that pertain to electronic warfare management and oversight within DOD, including those that relate to electronic warfare as a core or related capability of information operations. The table excludes, for example, responsibilities related to electronic warfare training and intergovernmental coordination, and the entities assigned such responsibilities.

[a]These organizations may be assigned non-electronic warfare capabilities in the listed documents. Here we only include those responsibilities that are related to electronic warfare, including electronic warfare as a core capability or related capability of information operations.

[b]Figure only provides Strategic Command responsibilities unique from those assigned to all combatant commands.

[c]DOD Directive 3222.4 predates the creation of Strategic Command in 2002.

DOD May Face Challenges in Its Oversight of the Evolving Relationship of Electronic Warfare and Cyberspace Operations

DOD may face challenges in its oversight of electronic warfare because of the evolving relationship between electronic warfare and cyberspace operations, specifically computer network operations; both are information operations-related capabilities. According to DOD, to ensure all aspects of electronic warfare can be developed and integrated to achieve electromagnetic spectrum control, electronic warfare must be clearly and distinctly defined in its relationship to information operations (to include computer network operations) and the emerging domain of cyberspace. In the previous section, we noted that DOD's directives do not clearly define the roles and responsibilities for the oversight of electronic warfare in relation to the roles and responsibilities for information operations. The current DOD Directive 3600.01 does not clearly specify what responsibilities the Principal Staff Advisor has regarding the integration of information operations-related capabilities—specifically the integration of electronic warfare capabilities with computer network operations.[32]

Further, DOD's fiscal year 2011 electronic warfare strategy report to Congress, which delineated its electronic warfare strategy, stated that the strategy has two, often co-dependent capabilities: traditional electronic warfare and computer network attack, which is part of cyberspace operations. Moreover, according to DOD officials, the relationship between electronic warfare and cyberspace operations—including computer network attack—is still evolving, which is creating both new opportunities and challenges. There will be operations and capabilities that blur the lines between cyberspace operations and electronic warfare because of the continued expansion of wireless networking and the integration of computers and radio frequency communications. According to cognizant DOD officials, electronic warfare capabilities may permit use of the electromagnetic spectrum as a maneuver space for cyberspace operations. For example, electronic warfare capabilities may serve as a means of accessing otherwise inaccessible networks to conduct cyberspace operations; presenting new opportunities for offensive action as well as the need for defensive preparations.

Current DOD doctrine partially describes the relationship between electronic warfare and cyberspace operations. Specifically, current joint doctrine for electronic warfare, which was last updated in February 2012,

---

[32] DOD Directive 5143.01 *Under Secretary of Defense for Intelligence* (Washington, D.C.: Nov. 23, 2005) provides the responsibilities of the Under Secretary and restates the Principal Staff Advisor duties stated in DOD Directive 3600.01.

states that since cyberspace requires both wired and wireless links to transport information, both offensive and defensive cyberspace operations may require use of the electromagnetic spectrum for the enabling of effects in cyberspace. Due to the complementary nature and potential synergistic effects of electronic warfare and computer network operations, they must be coordinated to ensure they are applied to maximize effectiveness.[33] When wired access to a computer system is limited, electromagnetic access may be able to successfully penetrate the computer system. For example, use of an airborne weapons system to deliver malicious code into cyberspace via a wireless connection would be characterized as "electronic warfare-delivered computer network attack." In addition, the doctrine mentions that electronic warfare applications in support of homeland defense are critical to deter, detect, prevent, and defeat external threats such as cyberspace threats.

DOD has not yet published specific joint doctrine for cyberspace operations, as we previously reported.[34] We recommended, among other things, that DOD establish a time frame for deciding whether to proceed with a dedicated joint doctrine publication on cyberspace operations and update existing cyber-related joint doctrine.[35] DOD agreed and has drafted, but not yet issued, the joint doctrine for cyberspace operations. According to U.S. Cyber Command officials, it is unclear when the doctrine for cyberspace operations will be issued.

The military services also have recognized the evolving relationship between electronic warfare and cyberspace operations. For example, to address future challenges, the U.S. Army Training and Doctrine Command conducted an assessment on how the Army's future force will leverage cyberspace operations and found that the Army's current vocabulary—including terms such as computer network operations, electronic warfare, and information operations—will become increasingly inadequate. According to the Army, these terms are becoming outdated as the operational environment rapidly changes due to factors such as technologic convergence of computer and telecommunication networks, astonishing rates of technologic advancements, and the global

---

[33] Joint Publication 3-13.1.

[34] See GAO, *Defense Department Cyber Efforts, DOD Faces Challenges in Its Cyber Activities*, GAO-11-75 (Washington, D.C.: July 25, 2011).

[35] GAO-11-75.

proliferation of information and communications technology. According to a Navy official, the Navy recognizes the evolving relationship between electronic warfare and cyberspace operations and is moving toward defining that relationship. However, the Navy first is working to define the relationship between electronic warfare and electromagnetic spectrum operations. In addition, Air Force Instruction 10-706, *Electronic Warfare Operations*,[36] states that traditional electronic warfare capabilities are beginning to overlap with cyberspace areas, which is resulting in an increased number of emerging targets such as non-military leadership networks and positioning, navigation, and timing networks.

According to U.S. Cyber Command officials, it is important to understand how electronic warfare and cyberspace operations capabilities might be used in an operational setting. Such information could then inform the further development of doctrine. U.S. Cyber Command officials stated that they have participated in regular meetings with representatives from the military services, the National Security Agency, defense research laboratories, and others, to discuss the relationship of electronic warfare and cyberspace operations. Moreover, the Under Secretary for Acquisition, Technology, and Logistics, has established steering committees that are developing road maps for the Secretary of Defense's seven designated science and technology priority areas—one of which is cyberspace operations and another is electronic warfare.

Conclusions

DOD faces significant challenges in operating in an increasingly complex electromagnetic environment. Therefore, it is important that DOD develop a comprehensive strategy to ensure departmental components are able to integrate electronic warfare capabilities into all phases of military operations and maintain electromagnetic spectrum access and maneuverability. DOD would benefit from a strategy that includes implementing parties, roles, responsibilities, and performance measures, which can help ensure that entities are effectively supporting such objectives, and linking resources and investments to key activities necessary to meet strategic goals and priorities. In the absence of a strategy that fully addresses these and other key elements, the DOD components and military services responsible for implementing this

---

[36]Air Force Instruction, 10-706, *Electronic Warfare Operations* (Washington, D.C.: Nov. 3, 2010).

strategy, evaluating progress, and ensuring accountability may lack the guidance necessary to prioritize their activities and establish milestones that are necessary to achieve intended results within a reasonable time frame. Moreover, as a result, DOD may not be effectively managing its electronic warfare programs and activities or using its resources efficiently. For example, an effective strategy could help DOD reduce the potential for unnecessary overlap in the airborne electronic attack acquisition activities on which we have previously reported.

The military's increasing reliance on the electromagnetic spectrum—coupled with a fiscally constrained environment and critical gaps in electronic warfare management—highlights the need for an effective governance framework for managing and conducting oversight of the department's electronic warfare activities. The absence of such a framework can exacerbate management challenges, including those related to developing and implementing an effective strategy and coordinating activities among stakeholders. Without additional steps to define the purpose and activities of the JEMSCC, DOD lacks reasonable assurance that this center will provide effective departmentwide leadership for electronic warfare capabilities development and ensure the effective and efficient use of its resources. As we previously reported, DOD acknowledges a leadership void that makes it difficult to ascertain whether the current level of investment is optimally matched with the existing capability gaps. Leveraging resources and acquisition efforts across DOD—not just by sharing information, but through shared partnerships and investments—can simplify developmental efforts, improve interoperability among systems and combat forces, and could decrease future operating and support costs. Such successful outcomes can position the department to maximize the returns it gets on its electronic warfare investments. In addition, multiple organizations are involved with electronic warfare and outdated guidance regarding management and oversight may limit the effectiveness of their activities. Both the Under Secretary of Defense for Acquisition, Technology, and Logistics and the JEMSCC have been identified as the focal point for electronic warfare within the department, yet it is unclear what each organization's roles and responsibilities are in relation to one another. Further, each organization's management responsibilities related to future iterations of the electronic warfare strategy report to Congress and working with the military services to prioritize investments remain unclear. Updating electronic warfare directives and policy documents to clearly define oversight roles and responsibilities for electronic warfare—including any roles and responsibilities related to managing the relationship between electronic warfare and information operations or electronic warfare and cyberspace operations, specifically computer network operations—would help ensure that all aspects of electronic warfare can be developed and integrated to achieve electromagnetic spectrum control.

Recommendations for Executive Action

To improve DOD's management, oversight, and coordination of electronic warfare policy and programs, we recommend that the Secretary of Defense take the following three actions:

- Direct the Under Secretary of Defense for Acquisition, Technology, and Logistics, in coordination with the Under Secretary of Defense for Policy and Strategic Command, and others, as appropriate, to include at a minimum the following information in the fiscal years 2013 through 2015 strategy reports for electronic warfare:
  - Performance measures to guide implementation of the strategy and help ensure accountability. These could include milestones to track progress toward closing the 34 capability gaps identified by DOD studies.
  - Resources and investments necessary to implement the strategy, including those related to key activities, such as developing electronic warfare organizational structures and leadership.
  - The parties responsible for implementing the department's strategy, including specific roles and responsibilities.

- Direct the Commander of Strategic Command to define the objectives of the Joint Electromagnetic Spectrum Control Center and issue an implementation plan outlining major implementation tasks and timelines to measure progress.

- Direct the Under Secretary of Defense for Policy, in concert with the Under Secretary of Defense for Acquisition, Technology, and Logistics, as appropriate, to update key departmental guidance regarding electronic warfare—including DOD Directives 3222.4 (*Electronic Warfare and Command and Control Warfare Countermeasures*) and 3600.01 (*Information Operations*)—to clearly define oversight roles and responsibilities of and coordination among the Under Secretary of Defense for Policy; the Under Secretary of Defense for Acquisition, Technology, and Logistics; and the Joint Electromagnetic Spectrum Control Center. Additionally, the directives should clarify, as appropriate, the oversight roles and responsibilities for the integration of electronic warfare and cyberspace operations, specifically computer network operations.

Agency Comments and Our Evaluation

In written comments on a draft of this report, DOD partially concurred with our first recommendation and concurred with our other two recommendations. Regarding our recommendation that DOD include in future strategy reports for electronic warfare, at a minimum, information on (1) performance measures to guide implementation of the strategy, (2) resources and investments necessary to implement the strategy, and (3) parties responsible for implementing the strategy, the department stated that it continues to refine the annual strategy reports for electronic warfare and will expand upon resourcing plans and organization roles; however, the department stated that the strategy was not intended to be prescriptive with performance measures. As we have previously stated, the inclusion of performance measures can aid entities responsible for implementing DOD's electronic warfare strategy in establishing priorities and milestones to aid in achieving intended results within reasonable time frames. We also have noted that performance measures can enable more effective oversight and accountability as progress toward meeting a strategy's goals may be measured, thus helping to ensure the strategy's successful implementation. We therefore continue to believe this recommendation has merit.

DOD concurred with our remaining two recommendations that (1) the Commander of Strategic Command define the objectives of the JEMSCC and issue an implementation plan for the center and (2) DOD update key departmental guidance regarding electronic warfare. These steps, if implemented, will help to clarify the roles and responsibilities of electronic warfare management within the department and aid in the efficient and effective use of resources. DOD's written comments are reprinted in their entirety appendix III.

We are sending copies of this report to appropriate congressional committees; the Secretary of Defense; and the Commander, U.S. Strategic Command. In addition, this report will be available at no charge on GAO's web site at http://www.gao.gov.

If you or your staff have any questions about this report, please contact me at (202) 512-4523 or leporeb@gao.gov. Contact points for our offices of Congressional Relations and Public Affairs may be found on the last page of this report. Key contributors to this report are listed in appendix IV.

Brian J. Lepore
Director, Defense Capabilities
and Management

Appendix I: Scope and Methodology

To assess the extent to which DOD has developed a strategy to manage electronic warfare we evaluated DOD's fiscal year 2011 and 2012 electronic warfare strategy reports to Congress[1] against prior GAO work on strategic planning, that indentified six desirable characteristics of a strategy.[2] The characteristics GAO previously identified are: (1) purpose, scope, and methodology; (2) problem definition and risk assessment; (3) goals, subordinate objectives, activities, and performance measures; (4) resources, investments, and risk management; (5) organizational roles, responsibilities, and coordination; and (6) integration and implementation. While these characteristics were identified in our past work as desirable components of national-level strategies, we determined that they also are relevant to strategies of varying scopes, including defense strategies involving complex issues. For example, identifying organizational roles, responsibilities, and coordination mechanisms is key to allocating authority and responsibility for implementing a strategy. Further, goals, objectives, and performance measures provide concrete guidance for implementing a strategy, allowing implementing parties to establish priorities and milestones, and providing them with the flexibility necessary to pursue and achieve those results within a reasonable time frame. Full descriptions of these characteristics are contained in appendix II.

We determined that the strategy "addressed" a characteristic when it explicitly cited all elements of a characteristic, even if it lacked specificity and details and could thus be improved upon. The strategy "partially addressed" a characteristic when it explicitly cited some, but not all, elements of a characteristic. Within our designation of "partially addressed," there may be wide variation between a characteristic for which most of the elements were addressed and a characteristic for which few of the elements of the characteristic were addressed. The strategy "did not address" a characteristic when it did not explicitly cite or discuss any elements of a characteristic, and/or any implicit references were either too vague or general. To supplement this analysis and gain further insight into issues of strategic import, we also reviewed other

---

[1] Office of the Under Secretary of Defense for Acquisition, Technology, and Logistics, Report to the Congressional Defense Committees. *Electronic Warfare Strategy of the Department of Defense* (Washington, D.C.: October 2010). The fiscal year 2012 report is classified.

[2] GAO-04-408T.

relevant strategic planning documents—such DOD's *National Defense Strategy*,[3] *Strategic Spectrum Plan*,[4] and *Net-Centric Spectrum Management Strategy*[5]—and interviewed cognizant officials from organizations across the department, including the Office of the Under Secretary of Defense for Acquisition, Technology, and Logistics; U.S. Strategic Command; and the Joint Chiefs of Staff.

To determine the extent to which DOD has planned, organized, and implemented an effective governance structure to oversee its electronic warfare policy and programs and their relationship with cyberspace operations, we reviewed and analyzed relevant DOD policies, doctrine, plans, briefings, and studies. Specifically, to determine how DOD has allocated electronic warfare authorities and responsibilities across the department, we reviewed and analyzed DOD policy, including DOD Directive 3222.4, *Electronic Warfare and Command and Control Warfare Countermeasures*; [6] DOD Directive 3600.01, *Information Operations*;[7] and the Secretary of Defense's *Memorandum: Strategic Communication and Information Operations in the DOD*.[8] We also reviewed relevant joint doctrine publications, such as Joint Publications 3-13, *Information Operations*[9] and 3-13.1, *Electronic Warfare*;[10] plans, including the 2008 and 2011 *Unified Command Plans*;[11] strategic documents, such as DOD's

---

[3] Department of Defense, *National Defense Strategy* (Washington, D.C.: June 2008).

[4] Department of Defense, *Strategic Spectrum Plan* (Washington, D.C.: February 2008).

[5] Department of Defense Chief Information Officer, *Department of Defense Net-Centric Spectrum Management Strategy* (Washington, D.C.: Aug. 3, 2006).

[6] DOD Directive 3222.4, *Electronic Warfare and Command and Control Warfare Countermeasures* (Washington, D.C.: July 31, 1992, Incorporating Change 2, Jan. 28, 1994).

[7] DOD Directive 3600.01, *Information Operations* (Washington, D.C.: Aug. 14, 2006, Incorporating Change 1, May 23, 2011).

[8] Secretary of Defense, *Memorandum: Strategic Communication and Information Operations in the DOD* (Washington, D.C.: Jan. 25, 2011).

[9] Chairman, Joint Chiefs of Staff, Joint Publication 3-13, *Information Operations* (Washington, D.C.: Feb. 13, 2006).

[10] Chairman, Joint Chiefs of Staff, Joint Publication 3-13.1, *Electronic Warfare* (Washington, D.C.: Feb. 8, 2012).

[11] Department of Defense, *Unified Command Plan* (Washington, D.C.: Dec. 17, 2008) and Department of Defense, *Unified Command Plan* (Washington, D.C.: Apr. 6, 2011).

fiscal year 2011 and 2012 electronic warfare strategy reports to Congress;[12] and classified and unclassified briefings, and studies related to DOD's identification of and efforts to address electronic warfare capability gaps, including DOD's 2009 *Electronic Warfare Initial Capabilities Document.*[13] We also reviewed DOD and military service reports, plans, concepts of operation, and outside studies that discuss DOD's definitions of electronic warfare and cyberspace operations. In addition, we interviewed cognizant DOD officials to obtain information and perspectives regarding policy, management, and technical issues related to electronic warfare, information operations, electromagnetic spectrum control, and cyberspace operations.

In addressing both of our objectives, we obtained relevant documentation from and/or interviewed officials from the following DOD offices, combatant commands, military services, and combat support agencies:

- Office of the Under Secretary of Defense for Policy
- Office of the Under Secretary of Defense for Intelligence
- Office of the Under Secretary of Defense for Acquisition, Technology, and Logistics
- Office of the Assistant Secretary of Defense for Networks and Integration/DOD Chief Information Officer
- Joint Chiefs of Staff
- Combatant Commands
  - U.S. Cyber Command, Fort Meade, Maryland
  - U.S. Pacific Command, Camp H.M. Smith, Hawaii
  - U.S. Strategic Command, Offutt Air Force Base, Nebraska
    - Joint Electromagnetic Spectrum Control Center, Offutt Air Force Base, Nebraska
    - Joint Electronic Warfare Center, Lackland Air Force Base, Texas
- U.S. Army
  - Office of the Deputy Chief of Staff of the Army for Operations, Plans, and Training, Electronic Warfare Division

---

[12] Office of the Under Secretary of Defense for Acquisition, Technology, and Logistics, Report to the Congressional Defense Committees. *Electronic Warfare Strategy of the Department of Defense* (Washington, D.C.: October 2010). The fiscal year 2012 report is classified.

[13] Department of Defense, *Electronic Warfare Initial Capabilities Document Unclassified Extract* (Washington, D.C.: Sept. 22, 2009). The full version of this report is classified.

- Training and Doctrine Command, Combined Arms Center Electronic Warfare Proponent Office, Fort Leavenworth, Kansas
- U.S. Air Force—Electronic Warfare Division
- U.S. Marines Corps—Headquarters, Electronic Warfare Branch
- U.S. Navy
  - Office of the Deputy Chief of Naval Operations for Information Dominance Electronic and Cyber Warfare Division
  - Naval Sea Systems Command, Naval Surface Warfare Center, Crane, Indiana
  - Naval Sea Systems Command, Program Executive Office for Integrated Warfare Systems
  - Navy Fleet Forces Cyber Command, Fleet Electronic Warfare Center, Joint Expeditionary Base Little Creek-Fort Story, Virginia
- Combat Support Agencies
  - Defense Information Systems Agency—Defense Spectrum Organization
  - National Security Agency, Fort Meade, Maryland

We conducted this performance audit from July 2011 to July 2012 in accordance with generally accepted government auditing standards. Those standards require that we plan and perform the audit to obtain sufficient, appropriate evidence to provide a reasonable basis for our findings and conclusions based on our audit objectives. We believe that the evidence obtained provides a reasonable basis for our findings and conclusions based on our audit objectives.

Appendix II: Desirable Strategy Characteristics

We previously identified a set of desirable strategy characteristics to aid responsible parties in implementation, enhance the strategies' usefulness in resource and policy decisions, and to better ensure accountability.[1] Table 2 provides a brief description of each characteristic and its benefit.

Table 2: Summary of Desirable Characteristics for a Strategy, Their Description, and Benefit

Characteristic	Summary description	Benefit
Purpose, scope, and methodology	Addresses why the strategy was produced, the scope of its coverage, and the process by which it was developed.	A complete description of the purpose, scope, and methodology in a strategy could make the document more useful to the entities it is intended to guide, as well as to oversight organizations, such as Congress.
Problem definition and risk assessment	Addresses the particular problems and threats the strategy is directed toward.	Use of common definitions promotes more effective intergovernmental operations and more accurate monitoring of expenditures, thereby eliminating problematic concerns. Comprehensive assessments of vulnerabilities, including risk assessments, can help identify key factors external to an organization that can significantly affect that organization's attainment of its goals and objectives and can help identify risk potential if such problem areas are not effectively addressed.
Goals, subordinate objectives, activities, and performance measures	Addresses what the strategy is trying to achieve, steps to achieve those results, as well as the priorities, milestones, and performance measures to gauge results.	Better identification of priorities, milestones, and performance measures can aid implementing entities in achieving results in specific time frames—and could enable more effective oversight and accountability.
Resources, investments, and risk management	Addresses what the strategy will cost, the sources and types of resources and investments needed, and where resources and investments should be targeted based on balancing risk reductions with costs.	Guidance on resource, investment, and risk management could help implementing entities allocate resources and investments according to priorities and constraints, track costs and performance, and shift such investments and resources as appropriate. Such guidance could also assist organizations in developing more effective programs to stimulate desired investments and leverage finite resources.
Organizational roles, responsibilities, and coordination	Addresses who will be implementing the strategy, what their roles will be compared to others, and mechanisms for them to coordinate their efforts.	Inclusion of this characteristic in a strategy could be useful to organizations and other stakeholders in fostering coordination and clarifying specific roles, particularly where there is overlap, and thus enhancing both implementation and accountability.
Integration and implementation	Addresses how a strategy relates to other strategies' goals, objectives and activities, and to subordinate levels of government and their plans to implement the strategy.	Information on this characteristic in a strategy could build on the aforementioned organizational roles and responsibilities—and thus further clarify the relationships between various implementing entities. This, in turn, could foster effective implementation and accountability.

Source: GAO.

---

[1] GAO, *Combating Terrorism: Evaluation of Selected Characteristics in National Strategies Related to Terrorism*, GAO-04-408T (Washington, D.C.: Feb. 3, 2004).

Appendix III: Comments from the Department of Defense

THE UNDER SECRETARY OF DEFENSE
2000 DEFENSE PENTAGON
WASHINGTON, D.C. 20301-2000

POLICY

JUN 2 3 2012

Mr. Brian J. Lepore
Director Defense Capabilities and Management
U.S. Government Accountability Office
441 G. Street, NW
Washington, DC 20548

Dear Mr. Lepore:

This is the Department of Defense response to the Government Accountability Office (GAO) draft report GAO-12-479, "ELECTRONIC WARFARE: DOD Actions Needed to Strengthen Management and Oversight," dated March 29, 2012 (GAO Code 351631). The Department partially concurs with the first recommendation and concurs with the other two recommendations contained in the draft report. With regard to the first recommendation, the Department continues to refine the annual strategy reports for electronic warfare to Congress and will expand upon resourcing plans and organizational roles; however, similar to the National Military Strategy, the Electronic Warfare Strategy is not intended to be prescriptive with performance measures, but rather to guide the analyses and processes the Department has established to balance risk and affordability that culminate in the annual budgetary and capabilities review process.

The Department appreciates the opportunity to respond to your draft report. We look forward to your continued cooperation and dialog toward our common goal of improving electronic warfare management and oversight throughout the Department of Defense. Should you have any questions, please contact Lt Col Patrick Eberle, (703) 697-3133, patrick.eberle@osd.mil.

Sincerely,

James N. Miller

Appendix IV: GAO Contact and Staff Acknowledgments

GAO Contact	Brian J. Lepore, Director, (202) 512-4523 or leporeb@gao.gov
Staff Acknowledgments	In addition to the contact named above, key contributors to this report were Davi M. D'Agostino, Director (retired); Mark A. Pross, Assistant Director; Carolynn Cavanaugh; Ryan D'Amore; Brent Helt; and Richard Powelson.

Related GAO Products

Airborne Electronic Attack: Achieving Mission Objectives Depends on Overcoming Acquisition Challenges. GAO-12-175. Washington, D.C.: March 29, 2012.

2012 Annual Report: Opportunities to Reduce Duplication, Overlap and Fragmentation, Achieve Savings, and Enhance Revenue. GAO-12-342SP. Washington, D.C.: February 28, 2012.

Defense Department Cyber Efforts: Definitions, Focal Point, and Methodology Needed for DOD to Develop Full-Spectrum Cyberspace Budget Estimates. GAO-11-695R. Washington, D.C.: July 29, 2011.

Defense Department Cyber Efforts: DOD Faces Challenges in Its Cyber Activities. GAO-11-75. Washington, D.C.: July 25, 2011.

Defense Department Cyber Efforts: More Detailed Guidance Needed to Ensure Military Services Develop Appropriate Cyberspace Capabilities. GAO-11-421. Washington, D.C.: May 20, 2011.

Defense Management: Perspectives on the Involvement of the Combatant Commands in the Development of Joint Requirements. GAO-11-527R. Washington, D.C.: May 20, 2011.

Electronic Warfare: Option of Upgrading Additional EA-6Bs Could Reduce Risk in Development of EA-18G. GAO-06-446. Washington, D.C.: April 26, 2006.

Electronic Warfare: Comprehensive Strategy Still Needed For Suppressing Enemy Air Defenses. GAO-03-51. Washington, D.C.: November 25, 2002.

Electronic Warfare: The Army Can Reduce Its Risk in Developing New Radar Countermeasures System. GAO-01-448. Washington, D.C.: April 30, 2001.

GAO's Mission	The Government Accountability Office, the audit, evaluation, and investigative arm of Congress, exists to support Congress in meeting its constitutional responsibilities and to help improve the performance and accountability of the federal government for the American people. GAO examines the use of public funds; evaluates federal programs and policies; and provides analyses, recommendations, and other assistance to help Congress make informed oversight, policy, and funding decisions. GAO's commitment to good government is reflected in its core values of accountability, integrity, and reliability.
Obtaining Copies of GAO Reports and Testimony	The fastest and easiest way to obtain copies of GAO documents at no cost is through GAO's website (www.gao.gov). Each weekday afternoon, GAO posts on its website newly released reports, testimony, and correspondence. To have GAO e-mail you a list of newly posted products, go to www.gao.gov and select "E-mail Updates."
Order by Phone	The price of each GAO publication reflects GAO's actual cost of production and distribution and depends on the number of pages in the publication and whether the publication is printed in color or black and white. Pricing and ordering information is posted on GAO's website, http://www.gao.gov/ordering.htm. Place orders by calling (202) 512-6000, toll free (866) 801-7077, or TDD (202) 512-2537. Orders may be paid for using American Express, Discover Card, MasterCard, Visa, check, or money order. Call for additional information.
Connect with GAO	Connect with GAO on Facebook, Flickr, Twitter, and YouTube. Subscribe to our RSS Feeds or E-mail Updates. Listen to our Podcasts. Visit GAO on the web at www.gao.gov.
To Report Fraud, Waste, and Abuse in Federal Programs	Contact: Website: www.gao.gov/fraudnet/fraudnet.htm E-mail: fraudnet@gao.gov Automated answering system: (800) 424-5454 or (202) 512-7470
Congressional Relations	Katherine Siggerud, Managing Director, siggerudk@gao.gov, (202) 512-4400, U.S. Government Accountability Office, 441 G Street NW, Room 7125, Washington, DC 20548
Public Affairs	Chuck Young, Managing Director, youngc1@gao.gov, (202) 512-4800 U.S. Government Accountability Office, 441 G Street NW, Room 7149 Washington, DC 20548

Please Print on Recycled Paper.

www.ingramcontent.com/pod-product-compliance
Lightning Source LLC
Chambersburg PA
CBHW081357170526
45166CB00010B/3122